The KidHaven Science Library

The Big Bang

by Don Nardo

KIDHAVEN PRESS
An imprint of Thomson Gale, a part of The Thomson Corporation

THOMSON
GALE

Detroit • New York • San Francisco • San Diego • New Haven, Conn. • Waterville, Maine • London • Munich

For more information, contact
KidHaven Press
27500 Drake Rd.
Farmington Hills, MI 48331-3535
Or you can visit our Internet site at http://www.gale.com

LIBRARY OF CONGRESS CATALOGING-IN-PUBLICATION DATA

Nardo, Don, 1947–
  The big bang / by Don Nardo.
    p. cm. — (KidHaven science library)
  Includes bibliographical references and index.
  ISBN 0-7377-2351-3 (hard cover: alk. paper)
  1. Big bang theory—Juvenile literature. I. Title. II. Series.
  QB991.B54N37 2005
  523.1'8—dc22

                                                                2004018734

Printed in the United States of America

Contents

Chapter 1

The Birth of Space and Time

Once, very long ago, the stars and planets, and all of the living things on Earth (and maybe on other planets) did not exist. The **universe**, the name used today to describe space and everything in it, had not yet been born. That means that space and time as we know them did not yet exist. No one knows what, if anything, did exist then. Perhaps there was only a void, a vast emptiness.

Whatever it was like before the birth of space and time, it was not destined to stay that way. Quite suddenly, about 14 or 15 billion years ago, there was a monstrous explosion. Its power was trillions of trillions of trillions times larger than that of the biggest nuclear bomb ever built by humans. The **Big Bang**, as scientists now call it, marked the birth of our universe. Matter, the atoms and other materials of which everything is made, rushed outward in all directions. Swiftly the universe expanded. And evidence collected by astronomers (scientists who study the wonders of the universe) shows that it is still expanding today.

Scientists wish to know how the Big Bang occurred. They especially want to figure out what happened in the seconds, minutes, and years immediately following the great explosion. It was during this critical early period that small particles of matter first

This computer-generated image shows how the Big Bang explosion might have looked.

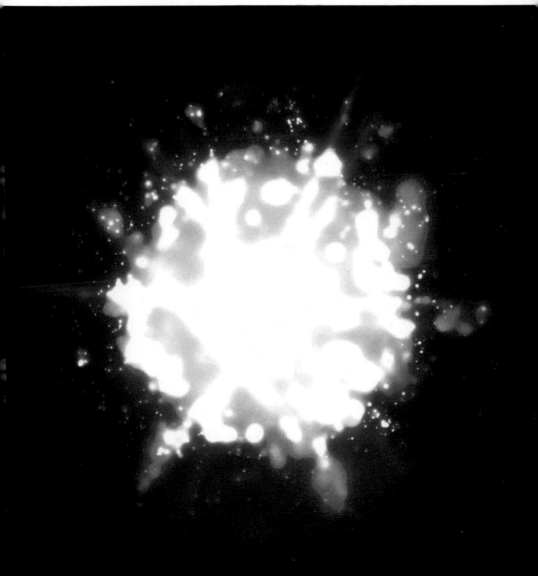

appeared. Those particles eventually became the building blocks of stars, **galaxies** (huge groups of stars), planets, and so forth.

But how can astronomers hope to discover what happened at the beginning of time? After all, the Big Bang took place in a distant part of space nearly 15 billion years ago. And no humans were there to witness and record what happened. To meet this challenge astronomers try to look at the universe's expansion in reverse. In a sense, it is like running a movie backward. For billions of years the galaxies and stars

In 2002 the Hubble Space Telescope (left) discovered young galaxies (below) that astronomers think were the building blocks of the universe.

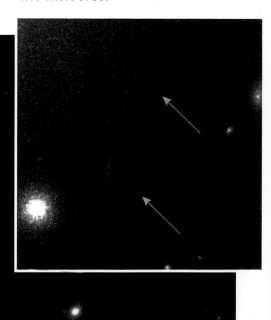

have been moving outward in all directions. Researchers attempt to trace this movement back as close as possible to its start. To do so, they use large telescopes, advanced satellites, complex mathematics, and powerful computers. They sometimes also make educated guesses.

The Expanding Bubble

Based on these studies, astronomers have pieced together a fascinating picture of the Big Bang and the earliest moments of the universe. First, all the materials that make up the present universe were squashed into an unbelievably tiny space. It was smaller than an atom. An atom is tremendously tiny by human standards. About 250 million average size atoms placed side by side would stretch just 1 inch (2.54 cm)! Astronomers call this tiny pre-universe mass a **singularity**. They also sometimes use a more colorful term—the "**cosmic** egg."

Exactly why this cosmic egg exploded is unknown. Perhaps it long existed in a perfectly balanced state and then somehow became unstable. One theory is that **gravity** (the natural force that causes each piece of matter to be attracted to others) did not yet exist. Rather, there was a force that was opposite to gravity—a sort of antigravity. Instead of attracting things, it pushed them away, causing them to explode outward.

However it happened, the titanic blast of the Big Bang marked the beginning of time and space as

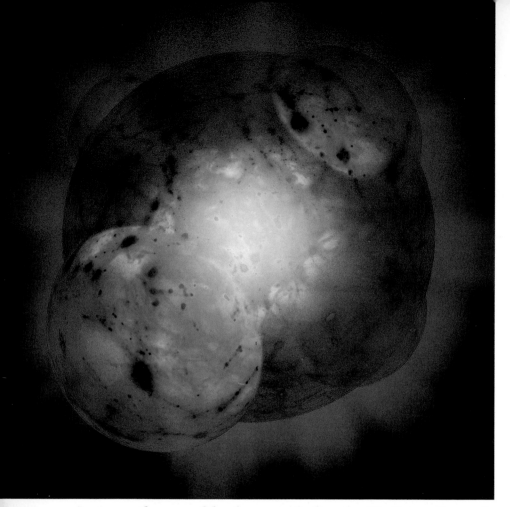

Just one-thousandth of a second after the Big Bang, the early universe expands into a large bubble.

humans know them. Indeed, space began expanding at an incredible rate. In far less than one-thousandth of a second, it doubled in size more than a hundred times. It ballooned from a microscopic speck smaller than an atom into a sphere, or bubble, 4,000 feet (1,219 m) wide.

That was only the beginning, for the cosmic bubble kept on expanding. At first it was unimaginably hot—an inferno measuring trillions of degrees. It was so hot, in fact, that ordinary matter as it is known today could not exist. There were no atoms. Instead, the bubble was filled with **subatomic particles**. These are particles thousands or millions of times smaller than atoms.

Among these particles were **electrons**, which are extremely small packages of energy. Scientists call many of the other tiny particles **quarks**. It was still too hot for particles of light, called **photons**, to form. So strangely, the growing bubble of hot, primitive matter gave off no light.

A computer created this illustration of three quarks sticking together to form a neutron.

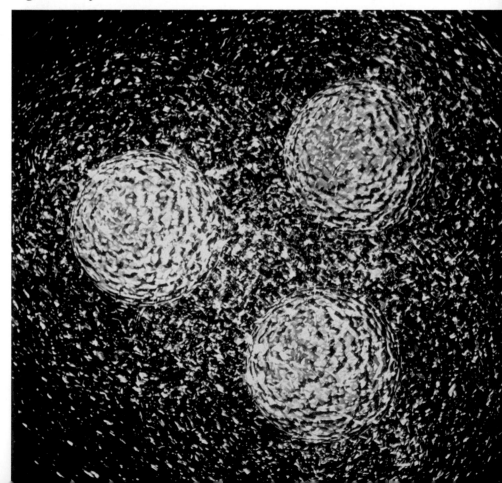

A Hot Soup of Particles

The dark, hot soup of quarks and other particles continued to expand at fantastic speeds. As it did so it began to cool somewhat. The cooling effect was very small at first. But it was just enough to allow some of the quarks to start sticking together. They began to form larger subatomic particles called protons and neutrons.

Everything described so far occurred in the very first second of the universe's existence. At the end of that initial second, the temperature of the cosmic bubble had dropped to perhaps 10 billion degrees Fahrenheit (5.5 billion degrees Celsius). Another second elapsed, and then a third. The seconds stretched into minutes, and all the while the cooling process continued.

At the three-minute mark, the temperature of the expanding universe was down to 1 billion degrees Fahrenheit (0.55 billion degrees Celsius). It was still much too hot to allow photons to escape as light. It was also too hot for full-fledged atoms to form. (Thanks to the heat, the free-flowing protons and electrons were too active to stick together and make atoms.) And the few that did manage to cling to one another quickly broke apart.

The Cosmic Infant

This situation changed, however, about three hundred thousand years after the Big Bang. This sounds

Structure of a Simple Atom

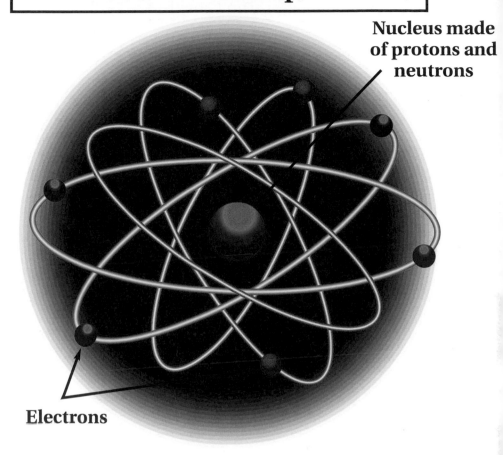

Nucleus made of protons and neutrons

Electrons

like a long time. And in human terms it certainly is, since human civilization is only a few thousand years old. Yet in cosmic terms, three hundred thousand years is like the blink of an eye. At that point the universe was still in its early infancy.

But the cosmic infant had reached a crucial turning point in its life cycle. Its immensely hot soup of particles had cooled to about 10,000 degrees Fahrenheit (5,538 degrees Celsius). Finally, conditions were right for the formation of true atoms.

The free-flowing protons combined with the free-flowing electrons to create the simplest atoms. When one proton combined with one electron, the result was an atom of hydrogen. Two protons and two electrons made an atom of helium. To this day these two elements—hydrogen and helium—remain the most common ones in the universe.

Another important event took place about three to four hundred thousand years after the Big Bang. Light and other kinds of radiation (streams of tiny, fast-moving particles) began to escape. As a result, the universe lit up and became very bright. There were no people around to see that light, however. The formation of planets and living things was still a long way off.

Chapter 2

Galaxies, Stars, and Planets Form

I n the first several million years that followed the Big Bang, no life of any kind existed in the universe. That universe still consisted of a rapidly expanding bubble of matter. The cosmic soup within the bubble contained large amounts of hydrogen and helium gases, along with numerous subatomic particles. It was impossible for life to spring from or exist in this environment. Living things are composed of highly complex clusters of atoms, or **molecules**. And such molecules did not yet exist.

Yet the young universe contained the seeds of the conditions that would one day give rise to life. First, as the expanding bubble got bigger and bigger, it continued to cool. When the universe reached the age of about 1 billion years, its temperature was down to about minus 200 degrees Fahrenheit (minus 129 degrees Celsius). Also, gravity, which had been created in the Big Bang, was beginning to affect matter in crucial ways. From this point on,

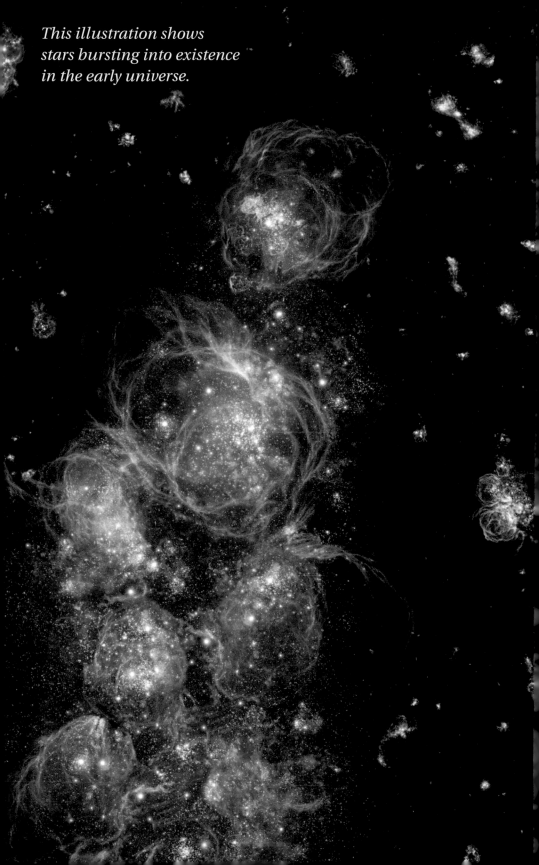

This illustration shows stars bursting into existence in the early universe.

gravity was destined to play a very important role in the universe. In time it would create the conditions and environments that would allow the rise of something truly special—life.

The First Stars

The first important job accomplished by gravity was to make clouds of gas. Little by little, tugged on gently by gravity, the atoms of hydrogen gas and helium gas were attracted to one another. As the ages passed, this caused large clouds of these gases to form. Many such gaseous clouds can still be seen today in various parts of the universe. Astronomers have photographed them and studied them very carefully.

In causing these gaseous clouds to form, gravity had just gotten started. Under its influence the atoms and primitive molecules within some of the clouds continued to move closer and closer together. And very slowly these clouds contracted, or shrank in size. At the same time, they became very **dense**, or compact, with more and more gases forced into a smaller and smaller space. This process gave off heat, which caused the gases in the clouds to warm up.

Thousands and millions of years passed. All the while, some of the gaseous clouds continued to grow smaller, more compact, and hotter. Astronomers are not completely sure what happened next. But their

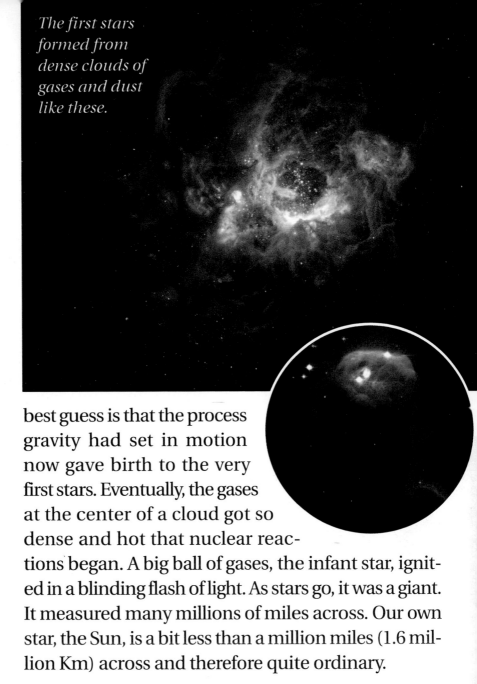

The first stars formed from dense clouds of gases and dust like these.

best guess is that the process gravity had set in motion now gave birth to the very first stars. Eventually, the gases at the center of a cloud got so dense and hot that nuclear reactions began. A big ball of gases, the infant star, ignited in a blinding flash of light. As stars go, it was a giant. It measured many millions of miles across. Our own star, the Sun, is a bit less than a million miles (1.6 million Km) across and therefore quite ordinary.

The Birth of Galaxies

For a while the big stars at the centers of the gaseous clouds were stable. For a few million years they continued to burn the large supplies of hydrogen that made up their great bulks. Meanwhile, in each of these clouds the remaining gases and dust swirled around the central star. Its immense gravity held these materials in its powerful grip.

Together a giant star and the materials orbiting it were the building blocks of a galaxy in the making. Today telescopes reveal a universe filled with galaxies—billions of them, in fact. Each is a swirling mass of hundreds of billions of stars. Earth and the other planets orbiting the Sun exist in the outer regions of one of these galaxies. Scientists call this galaxy the Milky Way.

Giant, irregular masses of gas begin to form galaxies in the wake of the Big Bang.

How did galaxies like the Milky Way form from the materials and forces unleashed by the Big Bang? First, the big stars at the centers of the clouds died off. They burned up their hydrogen and helium fuel extremely fast by cosmic standards. Eventually they became unstable. Gravity made their remaining materials collapse inward very violently. In most cases the result was a tiny, very dense kernel of matter that astronomers call a **black hole**. (Such an object is black because even light cannot escape its superpowerful gravity.)

The boxes show clusters of new stars that have formed near the center of a galaxy known as NGC 1365.

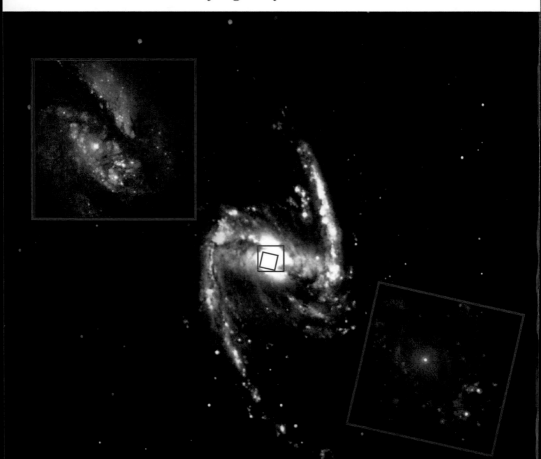

The enormous amounts of gases and dust that had been orbiting the star continued to move around the black hole. At the same time new stars were being born. The same process that had created the giant central star was making new, smaller stars throughout the gas cloud. Over and over again gravity caused patches of gas to contract and ignite into stars.

In time, a typical galaxy had many billions of stars of various sizes. Meanwhile, near the center of the galaxy, the black hole's gravity continued to attract matter. Stars that wandered too close to the black hole fell into it, feeding it and making it larger and more powerful. This series of events explains why today astronomers observe galaxies with billions of stars, each spinning around a common center. It also explains why most of these galaxies have giant black holes lurking at their centers.

Planets and Life

The grand process set in motion by the Big Bang did not end with the formation of galaxies. First, the tremendous force of that initial cosmic explosion caused the galaxies to continue moving outward in all directions. So the universe kept expanding in size.

Second, following the creation of billions of stars, gravity's job was not yet finished. Each young star had large amounts of dust and gases spinning

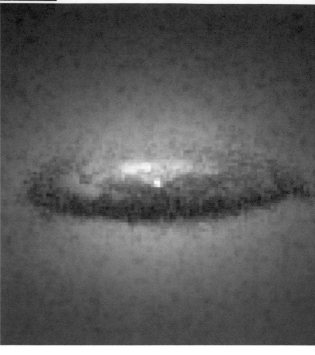

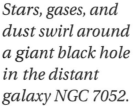
Stars, gases, and dust swirl around a giant black hole in the distant galaxy NGC 7052.

around it. These materials formed a wide, flat disk with the star at the center. The inner parts of the disk, those closest to the star, were hot. But the disk's outer sections were cooler. So they began to solidify, or harden, into small clumps of rock and ice.

As time went on, these clumps began to stick to one another, creating larger clumps. Some got to be thousands of feet across. Astronomers call these big clumps **planetesimals** (meaning planetary building blocks). The largest planetesimals had enough gravity to attract even more gas, dust, ice, and smaller planetesimals. Eventually, these grew into the planets, including Earth. (In some cases, slightly smaller chunks went into orbit around the planets, becoming their moons.)

It was on Earth (and probably other planets, too) that life, including humans, appeared and thrived. In a very real way, the Big Bang made life possible. The great ancient explosion gave birth to matter and gravity. In their turn, matter and gravity formed galaxies, stars, and planets. And on the planets, beings arose with enough intelligence and curiosity to figure out how their universe came to be.

Chapter 3

Discovering the Big Bang

The discovery of the Big Bang and the cosmic events it set in motion was one of the most important breakthroughs in the history of science. No single person made this breakthrough. Instead, it was a combined effort by many scientists working over the course of more than a century.

Improving technology was crucial in this effort. The evidence for the Big Bang had existed throughout recorded history. But for a long time human beings were unable to detect it because they lacked the proper tools. Among these tools are large telescopes that can see deep into space. Also important are instruments that can detect and measure faint radiation coming from space. Finally, satellites are crucial. They carry instruments above Earth's atmosphere. This is important because the atmosphere often distorts visual images and radiation coming from space.

Nebulae and the Size of the Universe

The first visual images in the trail of clues that led to the discovery of the Big Bang appeared in telescopes in the 1700s and 1800s. These were faint, blurry patches of light in the night sky. They were clearly not stars, which appear as sharp points of light. So astronomers called the foglike patches **nebulae**, from a Latin word meaning "mist."

Early astronomers referred to distant galaxies like this one as nebulae because of their foglike appearance.

The problem was that no one knew how far away the nebulae were. In those days most astronomers thought that the local galaxy, the Milky Way, made up the entire universe. In their view, therefore, the nebulae must be somewhere inside the Milky Way. Many experts suggested they were small gaseous clouds lying among or just beyond the planets in the Sun's own family (the Solar System).

In 1923, however, the mystery of the nebulae began to be solved. A huge new telescope, then the largest

Eighteenth-century astronomers at England's Greenwich Observatory study stars and nebulae.

The great astronomer Edwin Hubble poses beside a large telescope in the early 1950s.

in the world, went into operation on Mount Wilson, in California. American astronomer Edwin Hubble used this instrument to study the nebulae. He detected stars within many nebulae. Observations of these stars allowed him to measure the distances of some nebulae. He found that they were very far away.

Several of the nebulae were located far beyond the Milky Way. Hubble and other astronomers were astounded and excited when they realized that these distant nebulae were actually other galaxies. This meant that the universe was much larger than previously thought. In a single stroke the size of the known universe had increased by more than a hundred times.

The Expanding Universe

In the years that followed, Hubble and other astronomers closely examined the distant galaxies. It became clear that these objects were not standing

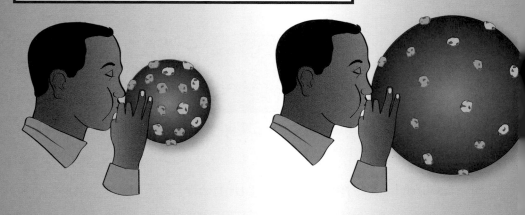

The Universe Expands

Like the spots on this balloon, all of the galaxies in the universe are moving away from each other as the universe expands.

still. Instead, they were moving rapidly through space. Not only that, they were all moving away from the Milky Way. Hubble concluded that this could mean only one thing. The universe must be expanding.

At first glance it seemed as though the Milky Way might lie at the center of the universe. After all, the other galaxies seemed to be racing away from it in all directions. But Hubble and other astronomers showed that this is only an illusion. No matter which galaxy one lives in, it will appear that the other galaxies are moving away from it.

Experts sometimes use the comparison of a raisin cake that someone places in an oven. The cake represents the whole universe and the raisins the galaxies inside it. The oven's heat makes the cake expand in size and the raisins begin moving away from one another in all directions.

Lemaître and the Cosmic Egg

The discovery that the universe is expanding was important for two reasons. First, it meant that astronomers might have a way to determine the universe's approximate age. They knew the rate at which the galaxies are moving. By tracing that movement backward, perhaps they could estimate how long the expansion had been occurring.

Second, tracing the expanding galaxies backward might give some idea of what the universe was like in its infancy. How did it begin? Or did it even have a clear-cut beginning? Could the universe have been expanding forever, with no definite starting point in space and time?

One scientist who tried to answer these weighty questions was Belgian astronomer Georges-Henri Lemaître. In 1927 he suggested that all the material in the universe was once part of a single object. He called it the "cosmic egg." Long ago, he said, the object suddenly exploded with enormous violence. Matter flew outward in all directions, and some of it eventually formed into galaxies, stars, and planets.

The problem for Lemaître was that he had no direct proof for his theory. Some astronomers, including Hubble, thought he was right. But others disagreed. They believed that the universe had no clear beginning and had existed forever. One in this group, American astronomer Fred Hoyle, poked fun at Lemaître's theory on a radio show in 1949. Hoyle called the supposed giant cosmic explosion the "Big Bang." And the name stuck.

Strange Radiation from Space

Hoyle and many other scientists continued to reject the Big Bang theory until 1965. In that year technology helped to provide firm evidence for the Big Bang at last. Using sensitive instruments, scientists Arno Penzias and Robert W. Wilson detected radio waves coming from the sky. Radio waves are a common kind of radiation people produce for various

Belgian priest and astronomer Georges-Henri Lemaître proposed the idea of the Big Bang in 1927.

forms of communication. But the radio waves Penzias and Wilson discovered were not made by humans. Strangely, they seemed to exist throughout the known reaches of space.

Penzias and Wilson realized that only one event could account for the radiation they had found. Scientists had long known that an explosion the size of the Big Bang would release huge amounts of a certain kind of radiation. That radiation would move outward as the universe expanded. And some of it would still be around as a sort of "fingerprint" of the great event. But they had never been able to detect this radiation.

Penzias and Wilson were thrilled because they knew that the radio waves they had found were the fingerprint of the Big Bang. Scientists named it CBR (or CMBR), which stands for cosmic background

An image produced by the Cosmic Background Explorer shows some of the cosmic background radiation in the Milky Way.

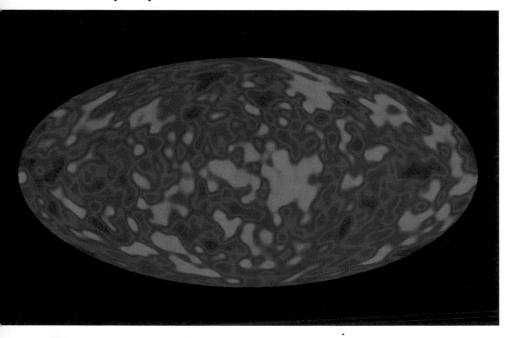

radiation. For their achievement Penzias and Wilson received the famous Nobel Prize. Since that time even more sensitive instruments have confirmed their findings. Among these is a satellite launched in 1989. It is called the Cosmic Background Explorer (COBE).

Georges-Henri Lemaître lived long enough to see his theory of the Big Bang proven correct in 1965. (He died the following year.) Today, most astronomers and other scientists accept the theory and view him as a person of great vision.

Chapter 4

The Fate of the Universe

Scientists have spent a lot of time and effort tracing the universe's expansion backward. In this way they have pieced together a convincing picture of the Big Bang and the dawn of space and time. But what about the universe's future? A number of astronomers have also tried to trace the universe's expansion forward in time. This allows them to speculate, or make educated guesses, about the

fate of both the universe and humanity. Will the universe go on expanding forever? Or will some force or forces act to create a different universal destiny?

The Big Crunch

One force that has already proven a major player in cosmic affairs is gravity. Gravity caused the formation of huge gas clouds in the universe. And under gravity's pull, these clouds contracted into galaxies, stars, and planets. Gravity also caused some stars to collapse into black holes. Large

Gaseous clouds contract to form stars (right); planets like the one below form around many of the stars.

black holes exist at the centers of many galaxies, growing ever larger by devouring more and more stars and planets.

Gravity will likely continue to influence the evolution of the universe. Astronomers have already seen gravity acting on a vast scale. Because the galaxies consist of billions of stars, they are very massive. And each galaxy has a very strong gravitational pull.

The Hubble Space Telescope photographed this image of two distant galaxies on a collision course.

As these huge objects move through space, from time to time some stray too close to one another. In such cases, gravity can cause two galaxies to crash together and merge. Many such collisions have been photographed. In addition, our own galaxy, the Milky Way, is presently absorbing a smaller galaxy that got too close.

Some astronomers suggest that in the future gravity will continue to pull on the galaxies. The gravity exerted by matter that floats between the galaxies and stars will also pull on the galaxies. All this gravitational pulling will slowly but surely slow down the universe's expansion. Over time that expansion will halt. Gravity will then make the galaxies start moving inward. Over the course of billions of years, they will race closer and closer toward a central point. Finally, all matter in the universe will merge into a tiny singularity similar to the one that gave birth to the Big Bang. Scientists call this event the "**Big Crunch**."

The Big Freeze

What will happen after the Big Crunch? No one knows for sure. But one possibility is another Big Bang. In other words, the new singularity might explode, creating another expanding universe.

Some astronomers have taken this idea a step further. Perhaps, they say, the new universe created in the Big Crunch will expand for many billions

This illustration shows what the Big Crunch might look like as gravity pulls galaxies inward toward a central point.

of years. Galaxies, stars, planets, and life will form, as they did before. Eventually, though, gravity's great hand will stop and reverse the expansion. A new Big Crunch will ensue, followed by still another Big Bang. On and on, the grand cosmic cycle will be repeated for all eternity. Because the word *oscillate* means to bounce back and forth, astronomers call this theory the "Oscillating Universe."

At present a number of scientists are doubtful that either the Big Crunch or Oscillating Universe will come to pass. The most recent measurements of the expanding galaxies show that most are moving apart at high speeds. These speeds may be great enough to overcome the pull of gravity.

In that case the fate of the universe will be very different from the Big Crunch. Over time, the galaxies will move farther and farther away from one another. At the same time, all of the matter in or between the galaxies will grow cooler and cooler. After many billions of years, even the stars will grow cold and dark. This theory is often called the "**Big Freeze**." Because it predicts a fatal leakage of heat, experts also refer to it as the "Heat Death" of the universe.

A dying star gives off eerie beams of light. It is possible that all stars will eventually die.

The Destiny of Life

If the Big Freeze does someday occur, and if humans or other intelligent beings still exist, what will these beings see? First, the galaxies will float far apart. Eventually, no other galaxies will be visible from any one

Intelligent beings living near the end of the universe may make their homes in space colonies like these.

Space Settlements

A Design Study

National
Aeronautics and
Space
Administration

galaxy. Beings in each galaxy will feel as though theirs is the only one in existence. Scientists call this lonely situation the "Big Rip."

What will happen to the last living things caught in the Heat Death and Big Rip? Will all life simply freeze to death? Or, if the Big Crunch occurs, will all life be crushed and fried to death?

Although these unpleasant results are possible, some scientists think there may someday be ways to avoid them. For example, researcher Freeman Dyson has studied the Big Freeze. He suggests that intelligent beings will be able to adapt to the changing conditions. The matter in the universe will slow down and grow colder. The bodies and brains of living things are made of matter. So they will slow down, too. This will take billions of years and life forms will not notice the gradual change. Each new generation will feel like they are the same as the last, even though they are a tiny bit different. In this way, humans and other intelligent beings can go on living and thinking forever.

But what about the fate of living things caught in the Big Crunch? Scientist and philosopher Frank J. Tipler believes this catastrophe can be avoided as well. In the far future, he says, humans will have very advanced technology. They will be able to merge their brains with computer-like machines. This new kind of intelligence will be able to think at a speed greater than that of the universe's collapse. So beings will *feel* like they still have plenty of time to exist and think, even when time is growing short. They might even be smart enough to figure out how to survive the Big Crunch.

These are only theories, of course. No one can say for sure what will happen billions or trillions of years from now. One thing is certain, however. The Big Bang universe gave rise to beings who can think, discover their own origins, and wonder about the future. They have a great thirst for knowledge and a strong instinct for survival. It seems likely that if there is a way to survive the universe's end, they will find it.

Glossary

Big Bang: The giant explosion that gave birth to the known universe.

Big Crunch: The theory that the universe will someday contract into a single, very dense mass similar to the one that gave birth to the Big Bang.

Big Freeze: The theory that the universe will continue to expand forever and grow colder and colder.

black hole: A superdense object with gravity so strong that not even light can escape it.

cosmic: Having to do with the cosmos (or universe).

dense: Highly compact.

electrons: Very small, highly energetic subatomic particles. Electrons combine with protons to form atoms.

galaxies: Gigantic groups of stars held together by their mutual gravities. Our galaxy is called the Milky Way.

gravity: A force exerted by an object that attracts other objects. The pull of Earth's gravity keeps rocks, people, and houses from floating away into space. It also holds the Moon in its orbit around Earth.

molecules: Combinations of two or more atoms.

nebulae: Clouds of gas floating in space or distant galaxies that look like gas clouds.

photons: Particles of light.

planetesimals: Small objects that orbited the early Sun and combined to form the planets.

quarks: Tiny subatomic particles that combine to make larger ones, including protons and neutrons.

singularity: A very small, dense, and massive point in space, such as the center of a black hole or the "cosmic egg" that exploded in the Big Bang.

subatomic particles: Tiny particles that are smaller than atoms and sometimes combine to form atoms.

universe: The sum total of all the space and matter that exists.

Books

Pam Beasant, *1000 Facts About Space.* New York: Kingfisher, 1992. An informative collection of basic facts about the stars, planets, asteroids, and other heavenly bodies.

Heather Couper et al., *Big Bang: The Story of the Universe.* London: Dorling Kindersley, 1997. A beautifully illustrated book that nicely sums up the Big Bang for basic readers.

Nigel Henbest, *DK Space Encyclopedia.* London: Dorling Kindersley, 1999. This critically acclaimed book is the best general source available for grade school readers about the wonders of space.

Jennifer Morgan, *Born with a Bang: The Universe Tells Our Cosmic Story.* Nevada City, CA: Dawn, 2002. A well-written introduction to the Big Bang, with the twist that the universe tells its own story.

Don Nardo, *Black Holes.* San Diego: KidHaven, 2003. Introduces the basics of gravity, stars, galaxies, and black holes.

Web Sites

The Big Bang: It Sure Was Big! (www.umich.edu/~gs265/bigbang.htm). A very well-written overview of the Big Bang that both students and general readers will find useful.

Galaxies (www.damtp.cam.ac.uk/user/gr/public/gal_home.html). An excellent introduction to galaxies, including what they are, how they formed, their structure, and so forth, with links to related topics.

The Hot Big Bang Model (www.damtp.cam.ac.uk/user/gr/public/bb_home.html). This introduction to the Big Bang was written by scientists but is easy to read and has several links to related topics.

Index

Picture Credits

Cover: PhotoDisc
ArSciMed/Photo Researchers, Inc., 9
© Bettmann/CORBIS, 24, 29
Mark Garlick/Photo Researchers, Inc., 36
Rick Guidice/NASA, 38 (below)
© Jupiterimages, 25, 33 (inset)
L5 News/NASA, 38 (inset)
NASA-GSFC, 6 (below), 14, 16 (below), 18, 20, 30
NASA-HQ-GRIN, 6 (circle inset), 16 (above), 34
NASA-MSFC, 23, 32-33, 37
Brandy Noon, 26
Ludek Pesek/Photo Researchers, Inc., 17
PhotoDisc, 5
Detlev Van Ravenswaay/Photo Researchers, Inc., 8
Royalty-Free/CORBIS, 11

In addition to his acclaimed volumes on ancient civilizations, historian Don Nardo has published several studies of modern scientific discoveries and phenomena. Among these are *The Extinction of the Dinosaurs, Cloning, Atoms,* volumes about the asteroids, black holes, the planet Pluto, and a biography of the noted scientist Charles Darwin. Mr. Nardo lives with his wife, Christine, in Massachusetts.

DATE DUE